ROYAL
OBSERVATORY
GREENWICH

Moons

Anna Gammon-Ross

Royal Observatory Greenwich
Illuminates

First published in 2022 by Royal Museums Greenwich, Park Row, Greenwich, London, SE10 9NF

ISBN: 978-1-906367-95-4

At the heart of the UNESCO World Heritage Site of Maritime Greenwich are the four world-class attractions of Royal Museums Greenwich – the National Maritime Museum, the Royal Observatory, the Queen's House and *Cutty Sark*.

rmg.co.uk

Typesetting by ePub KNOWHOW
Cover design by Ocky Murray
Diagrams by Dave Saunders
Printed and bound in Spain by Grafo

About the Author

Anna Gammon-Ross is a planetarium astronomer at Royal Observatory Greenwich. She's literally a rocket scientist, with a Masters degree in Space Exploration Systems from the University of Leicester, studying both human and robotic spacecraft engineering. If this book were written in the future, perhaps it would include a lunar space base similar to the one she helped design!

Entrance to the Royal Observatory, Greenwich, about 1860.

About Royal Observatory Greenwich

The historic Royal Observatory has stood atop Greenwich Hill since 1675 and documents over 800 years of astronomical observation and timekeeping. It is truly the home of space and time, with the world-famous Greenwich Meridian Line, awe-inspiring astronomy and the Peter Harrison Planetarium. The Royal Observatory is the perfect place to explore the Universe with the help of our very own team of astronomers. Find out more about the site, book a planetarium show, or join one of our workshops or courses online at rmg.co.uk.

Contents

Introduction

There are over 200 moons in our Solar System alone, from the one travelling so close to our planet that it is one of the easiest astronomical objects to spot in our skies, to those found towards the outer edges that are less well-known and harder to catch a glimpse of. We've been inspired to physically explore some, with space agencies all over the world planning both human and robotic missions to visit various moons, and astronomers and amateurs alike continue to make new discoveries that are enabling us to build up an ever more detailed picture of our celestial neighbourhood.

But before we start looking at any of them in particular, there's an important question that we first need to answer: what actually is a 'moon'?

Let's start by ruling out some things that are not essential to its definition. Starting at the beginning, a moon is not defined by how it formed. Many of the moons in our Solar System are thought to have formed with their planets, around 4 and a half billion years ago. It all started with a huge cloud of gas and dust that collapsed with most of its matter gathering towards the centre to form the Sun, the leftovers becoming a surrounding disc. The material within that disc then began to clump together, growing into bigger and bigger pieces until they formed planets, asteroids and dwarf planets. Many of those things were left with their own surrounding discs of gas and dust, which also stuck together to form pieces big enough to form many of the moons we find orbiting around them today.

Other moons, however, were originally not moons, but rather other celestial objects that flew a little too close to another object, getting captured by their gravity and pulled into orbit around them. We'll take a closer look at some possible examples of this as well as another moon formation method or two further on in this book, but, for now, just knowing that moons can form in a range of different ways is enough to count this out of our definition.

Nor is a moon defined by size. There's no particular size limit to being a moon. If we look at the objects in our Solar System, you can see there is a huge range – so much so that a few moons are bigger than some planets! That means there's no distinct cut-off point, no size when a moon becomes big enough to be a planet, for example, so we can't use that either.

One factor that is true of all moons, however, is that they are always less massive than their **primary**, the body they are orbiting. Although in everyday life the terms mass and size are often used interchangeably, in physics the two are different. Size is how big something is, its physical dimensions, whereas mass is how much 'stuff' or matter makes up something, so it's more linked to density and weight. For example, a watermelon would have more mass than a balloon of the same size, you can tell because it would feel heavier. Mass is also proportional to gravity. The more mass something has, the stronger the gravitational force it exerts and vice versa. This is not just true for things out there in space, but for everything – even you have mass, so you have your own gravity! It's just that your mass is *much* smaller than a planet or moon, so your gravity is too tiny to notice. The larger gravitational pulls of more massive objects dominate over the smaller gravity produced by less massive objects, so moons will always be pulled into orbit around another object with more mass, and therefore more gravity. This seems promising – we have finally identified something that is true for all moons! Unfortunately, this particular rule actually works for... well, everything! Planets are always less massive than the stars they orbit, stars are always less massive than the galactic centre they orbit, and so on. It's a universal rule related to gravity, rather than moons specifically.

A moon is not defined by its shape either. The majority of the most well-known moons are spherical, but this is simply because they are massive enough to achieve something called **hydrostatic equilibrium**. This might sound pretty complicated, but it's just a fancy name for quite a simple concept: when an object has enough mass and, therefore, gravity that it can overcome its own structural rigidity (that is the strength of the materials the body is made of), it forces itself to form the most compact shape possible and become roughly spherical. In our Solar System, we currently know of at least 19 moons that are massive enough to do this, but that's only a small fraction of the total moons discovered.

Neither does it matter what a moon is made of. We've discovered moons out there composed of many different types of rock or metal or ice. In theory, it may even be possible to have gaseous moons, though you would still need a significant size/mass difference between the moon and it's primary. Despite popular folklore though, we are yet to discover any moons made of cheese.

Luckily, there is an answer to our conundrum that is pretty straightforward: moons are astronomical objects that orbit planets or other star-orbiting objects. It is all simply to do with which object they travel around (see Figure 1). To be a little more scientific, another term that can be used instead of 'moon' is **natural satellite**. Usually, when we use the word 'satellite', we're talking about artificial satellites, human-made devices sent up into space to orbit around Earth. Around 5,000 artificial satellites are currently orbiting Earth, observing the planet, providing communication and navigation networks we can use on the ground, monitoring the weather, and much more. The most famous artificial satellite is, of course, the International Space Station (ISS). The first

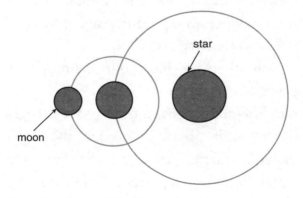

Figure 1: Moons orbit objects that orbit stars.

7

module of the station was launched in November 1998 and, since the year 2000, astronauts from all over the world have been living on it continuously.

'Satellite' on its own simply means anything that is orbiting another object, something that is true for all moons. Technically, Earth could be described as a satellite of the Sun, but what rules it out as a moon is the fact that the object it orbits is not a planet or a star-orbiting body, but a star. The word 'natural' is used in relation to astronomical bodies as they formed naturally, but also to make it clear we're talking about a different category of object to the artificial satellites.

Now that we have our definition in place, we're ready to start exploring all the different moons out there. We'll begin by taking a journey through the Solar System, starting close to the Sun and planet-hopping outwards. There's nothing to say moon-wise for the first two planets,

Mercury and Venus, however, as neither has any moons at all. There are a couple of reasons for this:

1. Mercury and Venus are both quite small planets with fairly low masses, so they don't have very much gravity. In general, the stronger an object's gravitational pull, the more moons it is going to be able to 'hold onto' or keep in orbit around it. For this reason, Mercury and Venus would struggle to pull any other objects into orbit around them, and it's why the gas giants, with much greater masses, have far more moons than the rocky planets.

2. These two planets are both quite close to the Sun. Gravity increases in strength the closer you are to the object producing it. An astronaut in a spacecraft above Earth would experience a much lower gravitational

pull than we do down on Earth's surface, for example. It also means that, because we are on Earth/really close to it, we are affected more by Earth's gravity than the Sun's, even though the Sun has significantly more mass and, therefore, gravity. As such, if an object was heading far enough into the centre of the Solar System to be affected by the gravities of Mercury or Venus, it would most likely be captured by the Sun's much greater gravity instead.

So, while it is possible for Mercury and Venus to have moons, it's very unlikely that an object would travel into the exact correct position for the gravitational pull to be right for them to become a permanent natural satellite.

Instead, let's start with...

Our Moon

The Moon is the only permanent natural satellite of our planet Earth and the fifth largest in the Solar System. It always seems a little confusing to call our Moon 'a moon' but there's a very good reason: it's simply that it was the very first moon we discovered! Our Moon is incredibly easy to spot, even without any kind of optical aid – you'll have seen it more times than you can recall. Because of this, for thousands of years, ever since our ancestors first looked up into the sky, we have known that our Moon exists. There are no other moons that can be seen from Earth so easily, so for a long time we thought that our Moon was the only one out there. It wasn't until telescopes were invented a few hundred years ago (more on that to come) that humans observed others orbiting around the other planets. By the time astronomers and other observers had the means to discover other moons, the name of our Moon had already stuck. Sometimes, though, the Moon is referred to by its Latin name 'Luna' instead, which gives us the adjective 'lunar'.

Lunar cycle and phases

The Moon orbits Earth approximately once every 27.3 days. It produces no light of its own, instead reflecting light from the Sun. Due to the trajectories of the orbits of the Moon and Earth, however, light from the Sun can't always reach the full side of the Moon we can see from Earth. The result is that we see different shapes or **phases** of the Moon.

The cycle of phases begins with a New Moon, when the side of the Moon facing Earth is fully in shadow with no sunlight able to reach it. This occurs when the Moon is on the same side of Earth as the Sun (see Figure 2).

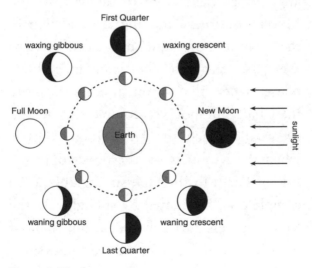

Figure 2: The lunar cycle.

As the Moon continues on its orbit, light from the Sun starts to illuminate the side we can see from Earth, giving us the waxing crescent phase (when a banana-like shape is visible in the sky). Once the nearside of the Moon is halfway illuminated, we have reached the First Quarter phase (so called because it occurs one quarter of the way through the full cycle). The Moon will also be at a right angle or perpendicular to the Sun. As the Moon continues moving around Earth, the amount of light it reflects grows. We pass through the waxing gibbous phase, when the Moon appears as more than half but not yet completely lit, and eventually reach the Full Moon. This is when the Moon is on the direct opposite side of Earth to the Sun, so it is able to be completely illuminated by sunlight. As the Moon travels through the second half of its orbit around Earth, we see those same phases again but in the reverse order:

waning gibbous, Last or Third Quarter, waning crescent and, finally, back to the New Moon again (see image 1). The New Moon and Full Moon phases are also the times in the cycle where solar and lunar eclipses occur respectively. We'll look more at eclipses later on in this book.

The full cycle of phases, from one New Moon to the next, takes around 29.5 days, which is a little bit longer than a single Moon-orbit of Earth. This is because Earth is also orbiting the Sun, so once the Moon has travelled around once, our planet has moved around a bit too, and the Moon has to go a little further to get back into the exact alignment with the Sun for a New Moon to occur again (see Figure 3).

Lunar features

Simply by observing the Moon with the naked eye, you can see that there are many

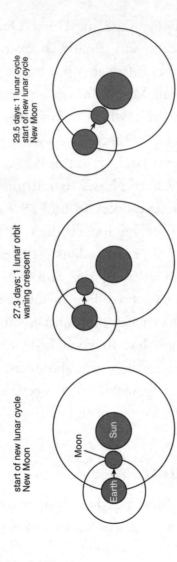

Figure 3: It takes the Moon approximately 27.3 days to orbit Earth, but the full cycle of lunar phases tales around 29.5 days because Earth is also moving around the Sun at the same time, so the Moon has to travel a little further to get into the exact alignment required for a New Moon to occur again.

16

features covering its surface. The large dark patches are regions called *maria*. This Latin term translates as 'seas' or 'oceans'. Long ago, when people were first looking up at the Moon, they tried to compare the things they could see with things they knew we had here on Earth – they thought there were literally huge seas covering the lunar surface, hence the name! However, there aren't any large bodies of water on our Moon. What we know now is that the dark regions observed are actually large plains of a different kind of rock. These features formed a few billion years ago, during a time when lots of space debris was colliding with the Moon's surface, creating huge impact basins. The basins have since been filled up and sort of smoothed over by lava and other lunar material to form the regions of **basalt**, a kind of dark, igneous rock. The largest lunar *maria* is called 'Oceanus Procellarum', which means 'ocean of storms'. With a surface area of roughly 4,000,000 km^2, it occupies around 10% of the total lunar surface.

The surrounding, lighter areas are known as the **lunar highlands**. As you may be able to guess from this name, these regions are much higher than the lunar *maria*. They consist of rock called **anorthosite**, another kind of igneous rock, formed when magma or lava creates intrusions inside existing rock. Anorthosite is more reflective than basalt, so the areas with a high concentration of the substance appear lighter and brighter than others when we look from here on Earth.

Of course, we can't talk about lunar features without mentioning craters, impact sites where smaller bodies have crashed into the Moon's surface and left a bit of dent behind. Craters aren't unique to our Moon, they can be found all over the Solar System – on planets, asteroids and other moons too. Our Moon appears littered with them, with over 9,000 named craters covering its surface. The largest among them appear as bright white spots to the naked eye.

If you look at images of the craters on the Moon, such as the region containing Marolycus and Moretus near the south pole on the Moon's visible side (see image 2), some are perfectly round, probably closer to what you imagine craters to look like, whereas others have more uneven, wonky edges, with mountain peaks at their centres. These differences are a consequence of how these craters formed. If an object collided with the Moon's surface with enough energy, it would heat up the rock into which it crashed enough to briefly melt it. The lunar surface would then quickly cool and resolidify after the landing. There is no way for the rock to maintain the higher temperature, so it would return to its normal state (at a rate that would likely be determined by the substances involved, ambient temperature at the crash site and the violence of the impact itself, among other things), but not before the rocky surface had been deformed enough to leave these irregular shapes behind. Any craters with imperfect shapes that you see, therefore, are the results of these highly energetic collisions.

Tidal locking

When looking at images of the Moon during its different phases, you will be able to make out many of these lunar features. It's to be expected that the fuller the Moon, the more features are visible. But something you might also notice is that the features are in the same relative positions for every phase (see image 1). This is because we are always looking at the same side or face of the Moon. The reason for this is thanks to a phenomenon known as **tidal locking**, which means the Moon spins on its axis at the exact same rate as it orbits Earth.

So there is a near side of the Moon, the side we can always see from our planet, and a far side, which always faces away from us (see image 3). The far side is sometimes referred to as the 'dark side' of the Moon but this is misleading because it is only every so often that that face is in complete darkness. In fact, the far side sees equal amounts of sunlight and shadow as the near side. As the Sun is always shining on half of the Moon, whatever phase we can see on the near side, the opposite is occurring on the far side. For example, when we have a New Moon, the far side is completely illuminated with a Full Moon phase, when the near side is a crescent, the far side is a gibbous, and both sides would experience quarter phases at the same time.

As the far side is never visible from Earth, we only discovered relatively recently what it looks like. The Soviet probe *Luna 3* produced the first images of the previously unseen side in 1959. It returned 29 photographs of what it had seen to Earth, capturing roughly 70% of the far side. China's *Chang'e 4*, named after the Chinese goddess of the Moon, achieved the first soft landing on the lunar far side in 2019. Though images like these reveal some of the same features as the near side – lunar highlands, craters (including one of the largest impact regions in the Solar System, the South Pole-Aitken basin) and *maria* – the proportion of surface covered by *maria*, however, is much lower on the far side, giving it a distinctly different appearance. Lunar geologists are still trying to work out why this is the case and one of the *Chang'e* mission's objectives is to study the chemical composition of the Moon's surface, including potentially unearthing materials from the lunar crust.

Lunar exploration

Ultimately, what makes our Moon unique is that it is the only place in the entire Universe, other than Earth, where humans have ever set foot. The first was, of course, the *Apollo 11* astronaut Neil Armstrong, back in 1969. Armstrong's capsule landed in one of the lunar *maria*, *Mare Tranquillitatis* or the Sea of Tranquillity, which in turn gave its name to the landing site, *Tranquillity Base*. Armstrong, along with his crewmate Edwin 'Buzz' Aldrin, spent more than two hours walking around on the Moon.

Over the course of the full *Apollo* programme, 12 people walked on the lunar surface and over 380 kg of rock and soil samples were collected, samples that have helped us learn a lot about our closest celestial neighbour. This includes the fact that lunar rock is very similar to the rock that makes up our planet. Research has shown that it actually has the same isotopic signature, or ratios of different types of **isotopes**. This has helped us to deduce that our Moon was most likely once part of Earth, forcing astronomers to completely change their theories on how the Moon formed.

Previously, it had been thought that about 4 billion years ago, a Mars-sized body, often referred to as 'Theia', crashed into Earth. Debris remnants of Theia from this collision were theorised to have stayed orbiting our planet before eventually clumping together to form our Moon. If this were the case, however, then the *Apollo* samples wouldn't have shown that lunar rock is far more similar to our own – there's less than a 1% chance that Theia would have had the same isotopic signature as Earth. It has been theorised instead that this impact may have broken off a chunk of our planet, which then became the Moon. The Moon has enough mass that it would have been able to reshape itself, through hydrostatic equilibrium, into the spherical form we see today.

Temporary moons

Our planet Earth may have only one permanent satellite, but it sometimes also has mini, **temporary moons** too. These are objects that get captured by a planet's gravity, becoming a natural satellite, but their orbit is not stable and so the body will either fly off out into space again or collide with its primary. We had one such temporary moon orbiting Earth from around 2016–17 until about May 2020. The little asteroid was named 2020 CD3 and had a diameter of only about a metre. It is thought to have flown away into space after its brief time as one of our moons and is due to pass by Earth again in March 2044, though probably not at a close enough distance to be captured by our gravity, even temporarily, again.

Martian Moons

Mars, the fourth planet from the Sun, is doing slightly better than Earth when it comes to natural satellites, as it is home to two moons: Phobos and Deimos. Discovered in 1877 by the American astronomer Asaph Hall, the moons are named after the Greek mythological twin gods, who were sons of Ares, the god of war, the Roman equivalent of whom is Mars. Their names translate as 'fear' and 'dread', which makes them sound like pretty terrifying objects, but they are among the smaller moons of the Solar System and rather oddly shaped –

definitely not scary at all (see image 4). In fact, they're often described as 'space potatoes'!

These two moons have irregular shapes because they don't have enough mass and, therefore, gravity to achieve hydrostatic equilibrium and make themselves spherical. Their shape previously led to a popular theory that they were originally asteroids from the nearby asteroid belt that flew close enough to Mars to be captured by the red planet's gravity. The shape and path of these moons' orbits doesn't quite fit this model, though, so it has also been proposed that they may have been formed from pieces of Mars itself that had broken off when space debris had crashed into the planet a few billion years ago.

Phobos is the larger of the two moons, with a diameter of around 22.2 km. Its main surface feature is a large impact crater about 10 km in diameter called 'Stickney'.

This crater is named after (Chloe) Angeline Stickney Hall, the American mathematician who encouraged her astronomer husband, the aforementioned Asaph, to persevere with his search for moons orbiting Mars. The grooves visible on Phobos' surface are thought to have been formed by rocks flying out from the collision that formed the Stickney crater.

No known moon orbits closer to its primary than Phobos does. It has an average orbital distance of 9,376 km, which is only a little further than the width of Mars itself. This distance is also, slowly but surely, reducing. Phobos is approaching Mars at a rate of about 2 cm every year, gradually spiralling its way inwards. Within the next 30 to 50 million years, the moon will either crash into Mars or the planet's gravity will tear Phobos apart before it can reach the surface. If the latter happens, the broken fragments of the moon will spread out within its current orbit to form a ring around Mars, similar to those we see around other planets in our Solar System.

Deimos, on the other hand, orbits further away from Mars, at a distance of about 23,458 km on average. It's also the smaller of the Martian moons with a diameter of around 12.6 km, making it a little over half the size of Deimos. Deimos is also very small in mass, so much so that the moon's gravity would be too weak to pull you back to its surface if you were to even do a small standing jump on it – if you did visit Deimos and try this, you would float away into space!

Both Phobos and Deimos are some of the least reflective bodies in the Solar System, with albedos of only 0.071 and 0.068 respectively. **Albedo** is a measurement, on a scale of 0 to 1, of reflectivity in space. A material or object with an albedo of 1 reflects all light perfectly, while an object with an albedo of 0 is completely matte and dark. For example, fresh snow is excellent at reflecting sunlight and has a high albedo of around 0.9, which is why snowy landscapes can seem dazzlingly bright to our eyes, whereas charcoal, one of the darkest materials, has an albedo closer to 0.04. The surfaces of both Martian moons are covered mostly in carbon-rich regolith, a blanket of loose rock and dust, which is the reason for their low albedo-values and very dark appearance.

Despite their darkness, in the future, when humans eventually stand on the surface of Mars(!), they will be able to spot both moons in the sky with the naked eye, depending, of course, on where they are on the planet. Phobos and Deimos have orbits that align with the equator of Mars, so they would be easiest to see from there, appearing smaller in the sky the further north or south you travelled, before becoming entirely invisible by the time you reached either of the poles.

Just as when you view our Moon from here on Earth, whether you could see these two moons would also depend on when you were looking up – sometimes the moons may have orbited around to be on the far side of the planet, below the horizon. Deimos takes a little over 30 hours to complete one orbit of Mars. This is actually so close to the length of one Martian day (a little over 24½ hours) that this moon would appear to move incredibly slowly across the sky, taking a couple of Earth days to go from moonrise to moonset. Phobos, on the other hand, whizzes around the planet, taking only a little over 7½ hours to complete one orbit of the red planet. This means you would see it travelling across the sky about three times every day! Not only that but it would appear to rise in the west and set in the east – the opposite direction travelled by Deimos and our Moon – because Phobos's orbital speed is faster than the planet's rotation. In terms of apparent size, from the equator Phobos would look about a third as big as a Full Moon viewed from Earth. Being smaller and much more distant, Deimos would more closely resemble a bright star or planet instead.

The surface of Mars is also the only other planet in the Solar System from where you could view an eclipse. We have two main types of eclipse here on Earth; a solar eclipse, where light from the Sun is blocked from our view by the Moon (see Figure 4), and a lunar eclipse, where the Moon passes into Earth's shadow (see Figure 5). For an eclipse to occur, there needs to be an alignment of the Sun, Earth and Moon. But, due to the Earth and Moon's orbits not being perfectly circular or in the same plane, the exact conditions required for even partial eclipses to be visible only occur a few times per year. This means that most of the time when the Moon is on the same side of the Sun as the Earth, we just experience a normal New

Figure 4: A solar eclipse. Light from the Sun is blocked from our view by the Moon.

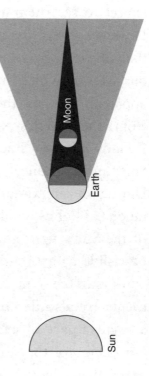

Figure 5: A lunar eclipse. The Moon passes into Earth's shadow.

33

Moon, and, usually, when the Earth is in the middle, just a Full Moon can be seen.

Eclipses on Mars occur far more frequently. As long as you are looking from the correct latitudes, you can expect to see them as a daily occurrence. However, they are much less impressive than those seen from our planet. Unlike our Moon, which has roughly the same apparent size as the Sun when viewed from the Earth's surface, Phobos and Deimos are too small to fully cover the Sun and produce total solar eclipses. During a maximum eclipse by these moons, Phobos could block up to 40% of the Sun's light and Deimos would just be visible as a tiny dot passing across the Sun's surface. As a consequence, these events are also described as transits, when a smaller part of a star's light is blocked out. Due to the composition of Mars's atmosphere, which differs from that of Earth, you wouldn't see any light at all on their surfaces during the equivalent of our lunar eclipse, they'd just vanish completely! For the more noticeable of the two, Phobos, both types of eclipse would only last a few seconds due to the moon's quick orbital speed. All these factors together mean we are certainly lucky to experience such impressively beautiful events on our planet, even if we have to wait a little longer to see them.

The Moons of Jupiter

Though we've covered the moons of half the planets in our Solar System, all four rocky or terrestrial planets, we've only accounted for three moons so far – just a tiny fraction of the 200+ moons in the Solar System. You may therefore have guessed already that the gas giants are home to the vast majority of these moons. This is partly due to the fact that they are much more massive planets – they are the gas *giants* after all! Even the smallest gas giant planet, Uranus, has a mass around 14½ times that of Earth, the largest of the 'rockies'. As we've established, this means the gas giants have stronger gravitational pulls and they are able to 'hold on' to larger numbers of natural satellites.

It also helps that these planets lie at a much greater distance from the Sun. Mars, the furthest rocky planet, is about 228 million km away on average, whereas the first gas giant, Jupiter, is on average approximately 778 million km from the Sun. The effect of the greater distances is that the gravitational pull from our central star is weaker, so it is less likely to be the more dominant force and unable to claim any would-be moons from those planets.

Galilean moons

At the time of writing, there are 79 known moons of Jupiter. The four largest were the first moons orbiting a planet other than Earth ever discovered. They are known as the **Galilean moons** because it was the Italian astronomer Galileo Galilei who first observed them back in 1610. The German astronomer Simon Marius also reportedly spotted these four moons independently around the same time, but as Marius didn't publish his observations, Galileo is credited for their discovery, which led to a bit of a rivalry between the two. It was, however, Marius who came up with the names we still use for these moons today: Io, Europa, Ganymede and Callisto. In mythology, these were all lovers of Zeus, the king of the gods and the Greek equivalent of the Ancient Roman god Jupiter.

Galileo was one of the first people to make a telescope and he used one of his homemade telescopes to make his observations of Jupiter (the main circle) and the Galilean moons (the surrounding stars) (see Figure 6). He wasted no time in publishing his observations, releasing them in a pamphlet called *Sidereus Nunicus,* meaning 'sidereal' or 'starry messenger', the same year he first observed the moons. By today's standards, the telescope Galileo used was pretty basic, but if you view these four moons through a pair of modern binoculars you will have a similar experience of what he observed in the 17th century. If you look at the right time, a couple of the moons are potentially even visible with the naked eye, though you would need very good eyesight to see them!

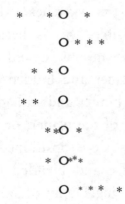

Figure 6: A replication of Galileo's observations of Jupiter's moons, first published in *Sidereus Nuncius* in 1610.

In Galileo's observations, you can see that the positions of the moons change relative to the planet from one night to the next, although they always appear in a single straight row, in line with Jupiter's equator. Some of these observations also indicate that not all four moons were visible. Due to this very specific apparent movement, this was the evidence that revealed to Galileo that what he was seeing were objects orbiting Jupiter, objects that could therefore be added to the 'moon' category. In the sketches where fewer moons are visible, this is because either: a) one of the moons would have been behind the planet and hidden from view; b) in front of Jupiter and so had merged in Galileo's view of the planet; or c) a couple of the moons were positioned too close to one another and blended together to appear as one single point of light.

Observing these movements for long enough reveals a pattern that can tell us how long each Galilean moon takes to complete one orbit. Io is the closest to Jupiter and, therefore, quickest, taking 1.8 Earth days. Next is Europa, which takes 3.6 days per orbit, then Ganymede with 7.2 days and, finally, the furthest of the four moons, Callisto, has an orbit time of 16.7 days. If you are particularly eagle-eyed, you may have already spotted that three of those numbers are connected. If you double Io's **orbital period** (the time it takes to orbit its parent) of 1.8 days, you get 3.6 days, one year on Europa. If you double it again, you get the time it takes Ganymede to orbit Jupiter, 7.2 days (see Figure 7). For every one orbit that Io completes, Europa travels halfway around the planet and Ganymede completes a quarter of its journey around Jupiter. They have a ratio of 1:2:4. When the orbit times of objects align like this we describe it as **orbital resonance**. This means that these three moons will align with each

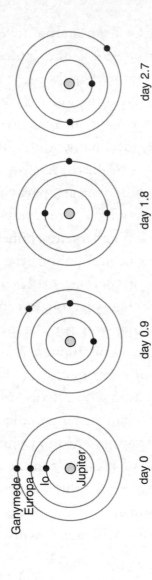

Figure 7: Three of Jupiter's Galilean moons are in orbital resonance. For every orbit of Jupiter that Io completes, Europa will have completed half its orbit and Ganymede a quarter.

other fairly often, so their combined gravities influence each other (and Jupiter) regularly. We'll look at some of the effects this has on the moons themselves a little later on.

In theory, given these regular orbit patterns, you could use these Galilean moons to help keep track of the time, like a natural clock. Galileo even used this idea as part of a proposal for determining longitude on Earth. While this method did work, it was quite complicated and required precise observations of the moons – not very practical for everyday life!

Nowadays, we not only have more advanced telescopes, but we have also launched spacecraft specifically to observe Jupiter and its moons. Between 1996 and 1997, the *Galileo* spacecraft captured detailed images of the four Galilean moons (see image 5) from which we have learnt a lot about the surfaces of these moons.

Ganymede

The largest Galilean moon is Ganymede. Earlier I said some moons are bigger than planets and here's our first example! Ganymede is the largest moon in the whole Solar System with a diameter of around 5,262 km, making it even bigger than the planet Mercury, which is approximately 4,879 km across. It's also not that far off the next planet up in size, Mars, which has a diameter of about 6,746 km. Ganymede is also the most massive moon in our Solar System (remember size and mass are different) at 148,185,846,875,052,000,000,000 kg. In fact, if it were orbiting the Sun instead of Jupiter, it would perhaps be considered a planet in its own right!

Ganymede is made up of three different layers: a metallic core, a surrounding shell of rock, and an icy outer layer (see Figure 8).

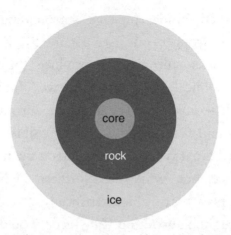

Figure 8: The internal structure of Ganymede. Its core is made of iron and surrounded by a layer of rock and then ice.

The metallic core is specifically made of iron, similar to Earth's core. This means that, like Earth, Ganymede has its own magnetic field – it's the only known moon to have one. This magnetic field was first detected by the *Galileo* probe and announced in 1996. Charged particles from the Sun are trapped in this field in a region surrounding the moon known as the **magnetosphere**. Ganymede's magnetosphere is very small because it is squeezed inside Jupiter's surrounding and much stronger one (see Figure 9). Despite this, it is still able to produce the same astronomical phenomenon we find on Earth – aurorae! The rings of aurora that surround Ganymede would be visible to the naked eye if you were standing on the moon's surface, but, while the Northern Lights on our planet are characteristically a greenish colour, the aurora on Ganymede would look red instead.

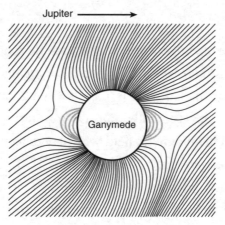

Figure 9: Ganymede has a very small magnetic field, which is squeezed in between the much stronger magnetic field of its primary, Jupiter.

Ganymede's average density is around 1.942 g/cm^3, which when combined with the densities of rock and ice, as well as the overall mass of the moon (roughly 1.482 x 10^{23} kg), tells us there must be a roughly equal ratio between the amounts of rock and ice found inside the moon. Not all the ice is thought to be solid, however. By studying Ganymede's aurora, in 2015 the Hubble Space Telescope was able to provide the strongest evidence yet that there is a huge subsurface salt-water ocean making up part of the moon's icy outer section. Jupiter's magnetosphere should cause Ganymede's aurora to move around a lot more than it appears to. Its limited effect could only be explained by there being something inside the moon itself that is more electrically conductive and so able to balance the planet's strong magnetic pull – a salty, buried ocean!

Ganymede's ocean is thought to hold around six times the amount of water than all of those on Earth combined, making it potentially the largest in the Solar System. Any discovery of liquid water in the Universe is very exciting because every kind of life that we know of here on Earth – every animal, every plant, even tiny bacteria – needs liquid water to survive. Any other places where water is found are therefore potential locations for finding alien life!

The surface of Ganymede, which has two main types of terrain, lies on top of the ice shell. Around 40% of the surface is darker and more heavily cratered, which shows the terrain is also older compared to other surface regions on this moon (we'll talk about why when we reach our next moon), with an estimated age of about 4 billion years. The remaining 60% of the surface is much lighter and covered in grooves and ridges, which are thought to have been formed by either volcanic or tectonic activity. Although these features are still ancient, they would have formed more recently than those found in the darker regions; so, though we know the lighter areas are younger, we aren't certain by how much.

Surrounding the moon is a very thin atmosphere. Despite being mostly oxygen, the ratios of the different types of gases, combined with how thin it is, mean it wouldn't be good to breathe, and, unlike Earth's protective atmosphere, it's not thick enough to keep Ganymede safe from things like solar radiation or meteorite debris. The different composition of the moon's atmosphere is also the reason why the aurorae here would appear red in colour – it is the Sun's charged particles interacting with these gases.

Callisto

The next Galilean moon out, and the second largest, is Callisto. Callisto has the most ancient surface in the whole

Solar System (including planets!) with an estimated age of around 4.5 billion years. You can determine the relative age of a solid planet or moon's surface by the number of impact craters on it. We know the average rate of space debris impacts in our Solar System, so by counting the number of craters an object has, you can calculate the approximate length of time that the body's outer layer has existed in its current form – the age of the surface. Callisto is the most heavily cratered body in the Solar System, so we know its surface must also be the oldest. Many astronomical bodies are able to refresh and change their surfaces through geological processes, such as tectonic plate movement or volcanic activity. Here on Earth, the planet's surface is also being reshaped and redeveloped by flowing water that erodes rock, different plants growing and moving topsoil, and even human activity – excavations change the ground so a new level becomes the 'surface', roads are resurfaced to add an artificial top layer – so the age of our planet's surface is actually very young.

Untouched by both human and plant life, and with no volcanic features, however, Callisto doesn't experience any of this – the rocky surface we see is the same one the moon was first formed with. In fact, there are so many craters on the surface of Callisto that the moon is reaching its saturation point. The vast majority of craters caused by new impacts will erase older ones rather than adding to the total number.

It's not just individual craters that appear on Callisto's surface but also chains of craters known as **catenae**, thought to be places where an object has been torn apart by a moon's gravity before colliding with it to form a line of impacts, and **multi-ring impact basins,** a series of concentric craters believed to be the site where an object has crashed, rebounded and collided again with the surface in roughly the same spot. The largest multi-ring crater in the Solar System can be found on Callisto. Named Valhalla, it consists of a bright central region with a diameter of around 360 km, with many surrounding rings of crater, the furthest of which has a diameter of about 3,800 km – greater than the distance between London and Cairo! Billions of years of impacts have also created fractures (places where the rocky surface has split or cracked) and scarps (steep slopes that form between two areas of different elevations, areas that have been fractured apart) on Callisto's surface, as well as deposits of icy material at higher locations and smooth dark material at lower points.

Io

My personal favourite of Jupiter's moons has got to be Io; there is nowhere else in the Solar System quite like it! Thanks to a combination of factors, its surface is constantly being refreshed and altered, more so than any other moon we know of. For starters, it is the closest moon to Jupiter so experiences the strongest effects from the planet's gravity. Add that to the orbital resonance discussed earlier and the effect is a moon pulled and pushed regularly, not just by Jupiter itself but also by the combined gravities of Europa and Ganymede in the opposite direction – it's as if it's been caught in the middle of a constant game of tug-of-war. All this force causes Io to deform and distort slightly, heating up the moon in a process known as **tidal heating**.

The heat produced has allowed Io to become the most geologically active astronomical body in the entire Solar System with an estimated 400 active volcanoes. These volcanoes have shaped virtually everything we can see on the surface of Io, creating features like lava flows and channels, as well as **paterae** – depressions with flat floors. The largest and strongest of these is called Loki Patera, which is over 200 km in diameter and contains an active lava lake surrounded by faults and mountains. There are no impact craters on Io, however. Any would-be craters caused by space debris colliding with the moon are filled back in by volcanic materials, smoothing the surface over again.

The colours we see on Io are caused by these volcanoes too. Mostly there's a yellowish colouring produced by a frosty coating of sulphur, but, if you look closely, you can also see reds, blacks, whites and greens too. These colours also mostly derived from different types and forms of sulphur.

The volcanic eruptions themselves are particularly impressive. The sulphur dioxide plumes produced on Io extend further into space than those on Earth. NASA's *New Horizons* space probe, for example, captured an image of a plume stretching around 330 km above the moon's surface, back in 2007. The impressive altitudes reached are due to weaker gravity pulling the volcanic clouds back towards Io, combined with a lack of atmosphere, which generates less air resistance. The plumes erupt so far into space and so frequently that Io is able to supply Jupiter's magnetosphere with as much as 1,000 kg of matter every second. This material goes on to form a cloud around the planet that helps to create Jupiter's aurorae.

Europa

The Jovian moon that is probably of most interest to the scientific community is Europa. The orbital resonance between the Galilean moons has affected this satellite too. Here, the tidal flexing and heating has allowed it to maintain a huge subsurface ocean of liquid, salty water around 100 m thick, beneath its frozen icy crust. In 2012, the Hubble Space Telescope captured an image of what appears to be a huge plume of water vapour from this ocean, shooting out from a location near Europa's south pole. The plume reached about 200 km in height from the surface of the moon – more than 20 times the height of Mount Everest! If these plumes do originate directly from the subsurface water, then eruption points like these would be excellent, natural access points for future spacecraft to explore Europa's oceans. It's something scientists are very keen to do as it looks to be one of the most promising places in the whole Solar System for some form of alien life to exist. Not only is there lots of liquid water, but also natural warmth from the tidal heating caused by Jupiter, Io and

Ganymede, as well as conditions suitable for naturally producing the right kinds of organic chemical ingredients, such as hydrogen and oxygen. In fact, some people have speculated that there could be huge sea creatures lurking beneath the surface of Europa, just waiting for us to discover them!

Above the icy crust, however, conditions are less than ideal. Europa has a very thin atmosphere of mostly molecular oxygen, not thick enough to provide any decent protection from the radiation found in space. There is so much radiation that if you were to spend even one Earth day on Europa's surface, you would either become severely ill or die.

Europa has the smoothest surface of any known object in the Solar System. The surface is too young and active to allow for the formation of craters or mountains – the plume material returns to the icy crust and refreezes, covering any features that try to form. Despite this, Europa's terrain is far from boring. There are round reddish spots known as **lenticulae** (Latin for 'freckles'), jumbled regions of cracks, ridges and plains known as **chaos terrain**, and maybe even ice spikes called **penitentes** at the equator, formed by sunlight heating the surface to create vertical cracks. The moon's most famous surface feature, however, has to be its 'tiger stripes', a series of crisscrossing dark lines or fractures known as **linae**.

Other moons

The remaining moons of Jupiter, those that aren't one of the four Galileans, are much smaller. They can mostly be split into a few different categories. First, there are the remaining 'regular' moons. These are known as the inner or Amalthea group, named after the fifth largest moon of Jupiter and the largest in this group,

Amalthea. The other three are Metis, Adrastea and Thebe. These moons are described as being regular because they have nice and neat orbit shapes and paths This makes them similar to the Galileans, but they are instead located closer to the planet. Each one is near enough that it takes less than one Jupiter day (around 10 hours) to complete a single orbit. Dust from the Amalthea group, produced by things like micrometeorite collisions with these moons, helps to maintain and replenish Jupiter's faint ring system.

The rest of Jupiter's moons are 'irregular', with orbits that are much more distant, less circular, and tilted at a much greater angle compared to Jupiter's equator. These differences between the regular and irregular moons' orbits shows that the regular satellites most likely formed at the same time and from the same materials that Jupiter itself did, whereas the irregular moons are more likely to be objects that formed elsewhere and have since been captured by Jupiter's gravity, pulled into orbit around the planet. The irregular moons can be split into four further sub-groups, the Himalia, Ananke, Carme and Pasiphae[1] groups, each named after the largest moon in their respective families and categorised mostly according to their distances from Jupiter.

---

[1] The moon Pasiphae was discovered in 1908 using images taken at the Royal Observatory, in Greenwich!

Saturnian Moons

Saturn is currently home to the most moons in the Solar System. At the time of writing, the total number of Saturnian moons was 83 (see Figure 10), the most recent discovery having been announced in November 2021: a moon with the catchy name 'S/2019 S 1', discovered by a research team from Canada–France–Hawaii Telescope (CFHT) observations. Prior to that came a discovery of 20 brand new moons, announced in October 2019. Clearly, the exact number changes frequently. This is true for the other gas giants too: the number of natural satellites discovered around Jupiter increased

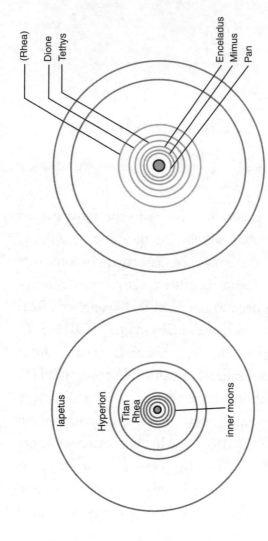

Figure 10: The orbits of a selection of Saturn's inner and outer moons.

by nine as recently as July 2018. So, it's fair to say that the numbers given in this book may be wrong by the time you are reading it!

The reason why we can't be sure if we have found them all yet is that most moons are really tricky to spot; the exceptions to this being those that we've already discovered! The more recently announced Saturnian moons are tiny, with each one less than 10 km in diameter – a distance the average person could walk in a couple of hours! As well as this, as these planetary satellites are a long way away from Earth, and because moons don't produce any light of their own like stars do, we have to rely on them being at the right angle (and made of the right materials to have a high enough albedo) to reflect sunlight in a way that makes them visible. As such, it's very easy to miss moons when studying the night sky and it's quite probable that we'll find more than 83 Saturnian satellites as we observe for longer and improve our technology.

The current tally also only accounts for moons that have confirmed orbits. Inside the rings of Saturn, there are hundreds of pieces of rock and ice orbiting the planet – that's essentially what Saturn's rings are made of. These pieces range in size from the microscopic (difficult to even detect, let alone count) to hundreds of metres across. The larger pieces are known as **ring moonlets**. Although, as we know, there is no formal size limit when it comes to defining moons, in the case of Saturn's rings, these particular satellites are commonly referred to as 'moonlets' because they are considered small compared to the other Saturnian moons, but large in comparison to the rest of the ring material.

Pan and Hyperion

A few satellites in and around Saturn's rings are massive enough to be able to use

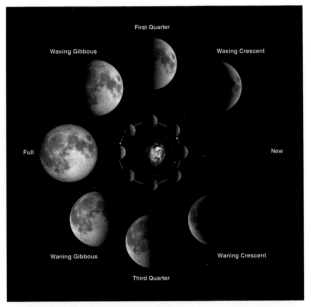

1. The changing view of the Moon seen from Earth. As the Moon orbits the Earth and the Earth orbits the Sun, the amount of sunlight reflected off the side of the Moon facing Earth changes and so we see different shapes, or 'phases' of the Moon. This full cycle takes 29.5 days. *NASA/Bill Dunford*

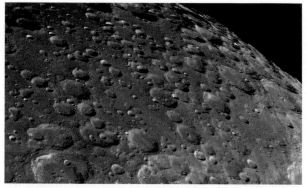

2. Craters, caused by the impact of a collision with a smaller object, are found on moons all over the Solar System. Our Moon has more than 9,000 named craters on its surface, including prominent craters such as Marolycus and Moretus, pictured here alongside countless much smaller ones.

Jordi Delpeix Borrell © National Maritime Museum, Greenwich, London. Courtesy of the artist.

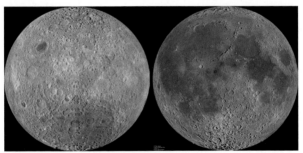

3. The Moon is tidally locked, so we see the same side of it from Earth at all times. Images and explorations of the far side of the Moon have revealed many of the same features, but a distinctly different appearance to the surface of the near side. Lunar geologists are working to uncover the reasons for this.

NASA/GSFC/Arizona State University

4. Phobos (top) and Deimos (bottom), the two moons of Mars, were discovered in 1877 by the American astronomer Asaph Hall. They are two of the smaller moons in the Solar System and their lack of mass means they are irregularly shaped. They are often referred to as 'space potatoes'. (Not to scale.)
NASA/JPL-Caltech/University of Arizona

5. The four largest moons of Jupiter are known as the Galilean moons, after Galileo Galilei, who first observed them in 1610. Clockwise from top left they are: Io, Europa, Callisto and Ganymede. With a diameter of 5,262 km, Ganymede is the largest moon in the Solar System and bigger than the smallest planet, Mercury. *NASA/JPL/DLR*

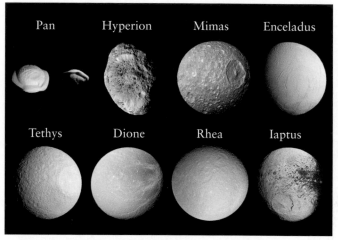

Pan Hyperion Mimas Enceladus

Tethys Dione Rhea Iaptus

6. At the time of writing, Saturn has the most moons in the Solar System, with 83, although new moons are frequently discovered orbiting Saturn and the other gas giants. Pan is one of several moons of Saturn known as a shepherd moon, as it is massive enough that its gravity helps to shape the rings surrounding the planet.

Top row, left to right: *NASA/JPL-Caltech/Space Science Institute, NASA/JPL/Space Science Institute, NASA/JPL-Caltech/Space Science Institute, NASA/JPL-Caltech/Space Science Institute*

Bottom row, left to right: *NASA/JPL-Caltech/Space Science Institute, NASA/JPL/Space Science Institute, NASA/JPL/Space Science Institute, NASA/JPL/Space Science Institute*

7. Titan is Saturn's largest moon and the second largest in the Solar System. It is the only moon known to have a significant atmosphere, and is covered in thick, orange clouds mostly made of nitrogen gas (left). In 2004, the *Cassini* spacecraft imaged Titan in infrared, allowing its surface to be seen for the first time (right). The images revealed lakes of liquid hydrocarbons near the moon's poles.

NASA/JPL-Caltech/Space Science Institute, NASA/JPL/ University of Arizona/University of Idaho

8. Uniquely among the planets in the Solar System, the moons of Uranus are all named after characters from literature. 27 moons have been identified so far, including the five largest pictured here, from left to right: Miranda, Ariel, Umbriel, Titania and Oberon. The grey areas represent regions of the moons that have not yet been imaged. *NASA/JPL*

their gravity to shape the rings themselves. They're known as **shepherd moons** because they manoeuvre the ring material like shepherds herding sheep. Some of these shepherd moons, such as Saturn's innermost proper moon, Pan, have used their gravity to clear gaps in Saturn's rings. Pan has one of the most unusual shapes in the Solar System, often described as being 'walnut-shaped' or looking like a 'space empanada' or even a 'cosmic ravioli'! The ridge you can see running around the circumference of the moon is formed from material from Saturn's rings that has been gathered up by Pan as it orbits inside them.

Pan is not the only strangely shaped moon of Saturn (see image 6). Hyperion was the first non-round moon ever discovered. Back in 1848, both the English astronomer William Lassell and a pair of American astronomers, William Cranch Bond and his son George Phillips Bond, independently spotted this moon and all three men have been given credit for finding it. We don't know the reason for its chaotic, sponge-like appearance but one theory is that it is a leftover fragment of a larger celestial body, perhaps even another moon, that was torn apart by an ancient collision.

Mimas

In total, only seven moons of Saturn are gravitationally rounded or in hydrostatic equilibrium. They can be split into two groups: four inner moons – those found within Saturn's extremely wide, second outermost ring, known as the E Ring – and another three found outside of it. The smallest and least massive of the inner round moons is Mimas. Although it is spherical, it still has a very distinctive shape. Thanks to its huge crater (named after the German-born, British astronomer William Herschel who discovered Mimas), the diameter of which is about a third of the moon's own diameter, this moon is often compared to a famous fictional space station fitted with a 'superlaser' from a certain 1977 sci-fi film – can you guess which one?![2]

Mimas is a little bit mysterious though. Despite the fact that it is in orbital resonance with some of the ring material and some of the other Saturnian moons (including Tethys, one of the other inner, rounded moons), and the gravities pushing and pulling on it mean that it should experience strong tidal forces and heating, Mimas actually shows no signs of past or present internal geologic activity. It is thought that this could possibly be due to its tiny size – the surface area of this moon is a little smaller than the land area of Spain, after all!

---

[2] Answer: Mimas is often compared to the Death Star from *Star Wars*, specifically *Episode IV: A New Hope*.

Enceladus

The next largest rounded moon is Enceladus. It's a moon that is, in a lot of ways, very similar to Jupiter's moon Europa; it also experiences orbital resonance (from a moon named Dione that we will talk about soon), and subsequently tidal flexing and heating to its interior, helping to maintain a large, subsurface ocean. This ocean of liquid water is about 10 km thick under the moon's icy crust, and, like Europa's, contains most of the chemical elements needed to support life. One difference between the two moons, however, is that Enceladus has volcanic activity. Not volcanoes like those we have on Earth, which erupt when molten rock (magma) from the planet's interior is forced upwards and breaks through the surface. Enceladus is too cold for magma. Instead, water from the subsurface ocean is forced up through the moon's icy crust to form features known as **cryovolcanoes**. These cryovolcanoes shoot out geyser-like jets of water vapour and some organic molecules, such as molecular hydrogen, mostly from Enceladus' south polar region. Over 100 of these **geysers** have been identified, all of which contribute to the moon losing approximately 200 kg of its mass every second. Most of this icy material helps to feed Saturn's E Ring; the rest falls back down to the surface of Enceladus like snow.

This is probably a contributing factor in making Enceladus one of the brightest objects in the Solar System with an albedo of around 0.99. It's mostly covered in fresh, clean ice with several different types of terrain. Some areas are relatively young, due to resurfacing from Enceladus' volcanic activity, while other regions are older and more heavily impacted, with many craters being heavily degraded with prominent domed floors. One of these larger craters,

Dunyazad, along with two neighbouring craters, has been described as looking like a snowman. The satellite is also home to lots of features, most likely caused by tectonic activity: linear cracks, scarps, plains, ridges, grooved terrains similar to those seen on Ganymede but a little more complex, and four 'tiger stripes' similar to Europa's.

Tethys

The next inner, round satellite of Saturn is Tethys. This moon has the lowest density of any discovered in the Solar System to date – being around 0.97 times as dense as liquid water! This tells us that Tethys must be almost entirely made of **water ice** with just a little bit of rock. This is supported by the fact that this moon is also very reflective; it has a relatively high albedo and is the second brightest Saturnian satellite.

Tethys is home to two particularly prominent surface features. Firstly, it has one of the largest craters in the Solar System, Odysseus, which has a total surface area over twice the size of the Republic of Ireland and a diameter nearly two fifths of the moon's whole surface. A crater this big should have been caused by a collision violent enough to shatter the moon so, as the moon is still intact, this event must have occurred when Tethys' interior was still molten and could withstand such an impact.

The other most noticeable feature on Tethys is a huge valley or **graben**, named Ithaca Chasma. It runs along the surface of Tethys for over 2,000 km, stretching from its north pole to its south pole. It is thought that Ithaca Chasma was either formed by the same force that produced the Odysseus crater or separately as a result of liquid water freezing and expanding under the already solid surface, causing it to crack.

Tethys also has two **co-orbital moons**: Telesto and Calypso. All three of these

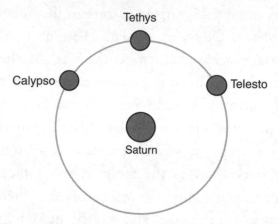

Figure 11: Tethys, Calypso and Telesto orbit Saturn at roughly the same distance from the planet. They also maintain the same distance from each other because their gravities are balanced.

satellites orbit Saturn at roughly the same distance from the planet, at about 295,000 km out. Their gravities are balanced so they maintain the same distance from each other, travelling at the same speeds (see Figure 11).

Dione

The final inner round moon is Dione. This satellite is similar to Tethys but has more craters as it is further from Saturn and experiences less tidal heating from its primary. The tidal heating on Tethys allowed the moon to stay molten for longer and so some of the old craters have been resurfaced over time. More of the craters on Dione can be found on the moon's **trailing hemisphere**, the side that is facing away from the direction the moon is orbiting in, than elsewhere on the satellite's surface (see Figure 12).

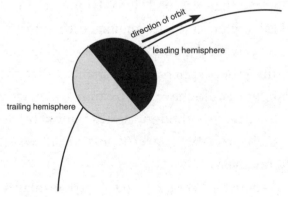

Figure 12: A moon's leading hemisphere is the one that faces the direction of travel. If a moon is tidally locked, the leading and trailing hemispheres never change.

As Dione is tidally locked (rotating at the same rate as it orbits Saturn, meaning that the same side is always facing towards the planet), the same half of the moon is always the trailing hemisphere. Logically, however, it should be that the side of the moon facing the direction of travel, the **leading hemisphere**, has more of the craters as it is the area that is travelling towards any potential impactors rather than away from them. Because of this anomaly, it is thought that one of the objects that crashed into Dione to form these craters actually managed to spin the moon around and switch the leading and trailing hemispheres. The chances that the satellite could have been spun by exactly 180 degrees though seems unlikely so there's probably more of this mystery still to uncover.

When the *Voyager* space probe imaged Dione back in 1980, it showed the moon to have 'wispy terrain' covering the trailing hemisphere. When the moon was revisited by the *Cassini* spacecraft in 2004, it became clear that these were actually bright ice cliffs and canyons, some of them stretching up several hundred metres. Dione has a slightly higher density than Tethys, indicating that about a third of this moon is a dense core most likely of silicate rock, the remaining two thirds being water ice. Like Tethys, Dione also has two co-orbital satellites, Helene and Polydeuces. These moons orbit at the same relative distances from each other as the Tethys co-orbital system.

Rhea

The first of the outer, round Saturnian moons is Rhea. It's very similar again to Dione and Tethys – so much so that they are often described as being like sister moons. All three are very cold with temperatures of around -170 °C on the sunlit side and -220 °C on the shaded side. These temperatures mean that the ice that is a substantial component of all three of these moons is so hard that it acts like rock. As Rhea is the furthest from Saturn of the three, it receives even less tidal heating than Dione so has even more craters. Like Dione, Rhea also has tall icy canyons. The walls of these canyons are very bright as any darker material falls off them to reveal fresh ice underneath.

Something unique to Rhea, however, is that this moon may have rings! In 2008, NASA announced a potential discovery of three narrow but relatively dense rings of particles orbiting Rhea. Unfortunately, no solid evidence has been found that they do exist, though a chain of bluish marks around the moon's equator looks as though it could have been formed by impacts from ring material. This implies that they may once have existed, even if they don't now.

Iapetus

The most distant round moon, Iapetus, was discovered by the Italian astronomer, Giovanni Domenico Cassini, in 1671. He also discovered Rhea the following year, as well as Tethys and Dione in 1684. Cassini noticed strange changes in the brightness of Iapetus as he observed it, which he correctly identified as being due to one hemisphere being covered in darker material than the other.

Iapetus, like many of the other Saturnian moons, is tidally locked with its parent – the same halves of the body are always fixed as the leading and trailing hemispheres. The leading hemisphere is a darker reddish-brown. Its albedo is very low with a value of around 0.03–0.05. The trailing side, on the other hand, is much brighter and more reflective with an albedo of 0.5–0.6. One of the more popular theories as to this colour difference, is that

Iapetus may be sweeping up dark material from another nearby moon, Phoebe. The surface of Phoebe is very dark, and it experiences lots of micrometeoroid impacts that kick up this dark dust and then go on to feed into the faint, most distant ring of Saturn, the Phoebe ring. This ring is very large and some of the material has moved inwards, widening it even further. It also moves around Saturn in the opposite direction to Iapetus, meaning that the moon ends up travelling straight into the ring material, splattering the leading hemisphere with the darker dust.

Titan

The final gravitationally rounded moon of Saturn, and my personal favourite, is Titan. This was the first Saturnian moon to be discovered back in 1655 and was originally named 'Saturni Luna', meaning 'moon of Saturn'. It was only after six more were discovered that it was renamed 'Titan' by the English mathematician and astronomer John Herschel, whose father, William, discovered both Mimas and Enceladus.

The Titans were a race of Greek mythological giants, some of whom lend their names to other Solar System objects, including Uranus, Hyperion, Tethys and Rhea. 'Titan' is also used frequently in popular culture for planets, moons and alien races, such as for the home world of the purple Marvel villain, Thanos.

The second largest moon in the Solar System and the largest of Saturn, Titan has a diameter of around 5,150 km, making it roughly one and a half times the size of Earth's moon and, like the largest moon Ganymede, greater in size than the planet Mercury. It is the most massive gas-giant moon relative to its planet and is roughly 3.4 times the size of Saturn's second largest moon, Rhea.

At first glance, nothing about Titan looks all that remarkable – it's a fairly smooth orange-coloured orb, with none of the dramatic shapes or features we've seen with previous moons. But what causes this plain appearance is actually something very exciting and unique: Titan is the only known moon to have a significant atmosphere. It's surface is covered by thick, orange clouds made mostly of nitrogen gas. Titan and Earth are the only two known places to have nitrogen-rich atmospheres. However, the air covering this moon is much denser than what we have here on Earth, meaning that pressure on the surface of Titan is nearly one and a half times that on Earth. Thanks to the density of these clouds, it was originally thought that Titan was much bigger, even bigger than Ganymede in fact!

Looking up from the moon's surface, through Titan's atmosphere, would be very different to our experience on Earth.

Due to the thick cloud cover, the sky would appear dark orange in colour and around 100 to 1,000 times dimmer than on our planet during the day. Just as it would be tricky to see anything up in the sky from Titan, it is difficult to see anything of its surface when looking from above. When learning more about it, some different methods, rather than just straight observation, need to be used.

In 2004, the spacecraft *Cassini* reached Saturn with the aim of studying the planet and its moons. It imaged Titan in **infrared**, a type of light that we can't see with our eyes, allowing the first ever images of the moon's surface to be seen (see image 7). Data produced by *Cassini* revealed that the surface of Titan is relatively flat. The tallest peak reaches only 3.3 km and, generally, most mountains are 50 m or less in altitude (for perspective, Mount Everest has an elevation of around 8.8 km). Most excitingly, *Cassini* mapped lakes towards the polar regions of Titan. These ranged in size from 1 to 100 km across. This makes this moon the only other known object in the Universe other than Earth to have stable bodies of liquid on its surface. Titan's lakes aren't made of water, but of liquid hydrocarbons instead. Despite this difference it is still a very important discovery and something else that makes Titan a unique place.

The *Cassini* spacecraft also carried on board a lander, *Huygens*, named after the Dutch astronomer who discovered Titan. This spacecraft was designed to land on the moon's surface. As it parachuted down through the thick atmosphere, *Huygens* captured images of Titan's terrain as we would see it with our eyes, or in visible light, revealing that many of the surface features appear to have been formed by past surface liquid movement and helping us understand further what it would be like to visit this moon.

Moons of the Ice Giants

We're towards the outer edge of the Solar System now, having reached the moons of the final two planets, Uranus and Neptune. These planets are sometimes known as the ice giants because of their composition. Saturn and Jupiter are around 90% hydrogen and helium, whereas, while Uranus and Neptune do have an outer gassy layer, their main ingredient is a mantle made of ices. (When we talk about ices out here, we don't just mean water ice like you find in your freezers, but instead anything with a freezing point above about -173.15 °C, so that also includes substances like methane and ammonia.) This composition difference is due to the fact that these planets have lower masses compared to the gas giants and, therefore, they can't gravitationally hold on to as much hydrogen and helium. The term 'ice giant' is also apt because we are now so far away from the Solar System's main heat source, the Sun, that the temperatures are incredibly cold. The moons of the ice giants, just as distant from the Sun, are extremely cold and icy too.

Moons of Uranus

The moons of Uranus are named in a different way to those of any other planet. Most moons in our Solar System are named after gods and goddesses, primarily from ancient Greek and Roman mythology. All of Jupiter's moons are named after lovers or daughters of Jupiter or his Greek equivalent Zeus, for instance, and the satellites of Neptune are named after Greek water gods and deities (as the planet itself shares its name with the God of the sea). Uranus' moons, however, are named after characters from literature. They are mostly taken from the plays of Shakespeare but there are also a handful from 'The Rape of the Lock', a satirical poem published by Alexander Pope in the 18th century. It was John Herschel who named the first few moons specifically after airy spirits from these works – fitting as Uranus is the god of the sky – but subsequent moons were named after other characters.

The 27 moons of Uranus we've identified so far can be split into three categories: the inner moons closest to the planet, the large or major moons, and the irregular moons, which orbit at the furthest distance (see Figure 13).

The inner moons are tiny, chaotic and appear to be unstable. Their orbits cross

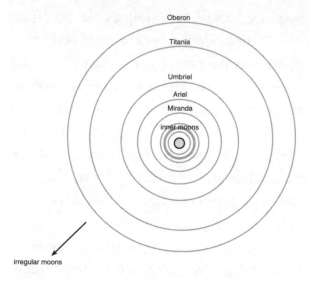

Figure 13: The orbits of the five major moons of Uranus. There are a number of smaller inner moons as well as irregular moons that lie beyond Oberon's orbit.

over each other, raising the possibility of collisions between them in the future, as well as being in and around Uranus' small ring system, such that they could be considered shepherd moons like those we find using their gravity to shape the rings of Saturn. All of these moons are very dark too, with low albedos, making them tricky to spot. It is thought that there are probably more moons in this group even closer to Uranus, but we haven't found them yet.

The five major moons of Uranus, in order of increasing distance from the planet, are: Miranda, Ariel, Umbriel, Titania and Oberon (see image 8). Up until NASA's *Voyager 2* spacecraft visited the Uranian system in 1986, these were the only moons of Uranus that we knew to exist.

Miranda may be the smallest of these five with a diameter only one seventh that of our Earth's Moon, but, in my opinion, it is the most interesting. It has a unique surface that looks like it has been smashed and stuck back together again but not quite correctly. This may actually be what happened to give Miranda its unusual appearance; a piece of space debris may have crashed into the moon, tearing it apart, before it managed to pull itself back together thanks to the mutual gravitational attraction of the pieces. Alternatively, the jigsaw-puzzle-like look may have been caused by subsurface ice melting and refreezing at various points in time, shifting the rocky parts around it, perhaps caused by collisions with things like micrometeorites.

Miranda's surface is one of the most varied in the Solar System, home to many different types of features such as ridges, valleys, craters, and scarps. There are canyons that are up to 12 times deeper than the Grand Canyon here on Earth, as well as the tallest cliff we know of in the whole Solar System, Verona Rupes, which is estimated to be around 20 km high. The cliff's height combined with Miranda's low gravity means that, if you were to drop a stone from the top of Verona Rupes, it would be a good ten minutes before it hit the ground at the bottom.

The next moon along, Ariel, is the brightest of the major Uranian moons, but that's not really saying much – it is still only able to reflect no more than a third of the sunlight it receives. The surfaces of all five of these moons seem to be darkened by some sort of carbonaceous material, which means mostly carbon compounds. Ariel's surface also seems to be the youngest in this satellite group. It is relatively smooth with a lower proportion of cratered areas than the surrounding moons.

Umbriel, the next moon out from Uranus, is the opposite to Ariel: it is the darkest of the five moons, reflecting only around 16% of the light that hits it, and its surface is ancient, covered in impact craters. The most striking feature of Umbriel is a mysterious bright ring, located at the centre of a crater named Wunda. This could either be material left from a recent collision or perhaps some ice from the moon itself that, having migrated across the surface, got trapped in the relatively cold conditions at the bottom of this crater.

Titania is the largest moon of Uranus and the eighth biggest in the whole Solar System, with a diameter of around 1,600 km. It makes up about 40% of the total mass of Uranian moons on its own. Its density shows that Titania, like all the large moons in this group, is made up of roughly equal parts water ice and silicate rock. Both Titania and the most distant of the major moons, Oberon, were discovered by William Herschel in 1787. (Herschel also discovered Uranus itself a few years earlier, as well as Saturn's moons Enceladus and Mimas, in 1789.) They are named after the king and queen of the fairies from Shakespeare's *A Midsummer Night's Dream*. Oberon's surface is the most ancient of all the large Uranian moons. It is covered in even more craters than Umbriel. The impacts of charged particles alongside this debris over time may have discoloured Oberon to produce its distinctive reddish hue. There are also lots of features known as **chasmata**, long, deep valleys or depressions with very steep sides.

The remaining moons are in the irregular group that lies beyond Oberon's orbit. The variation in their orbital shapes – some are more circular, others more elliptical – angles and even directions is most likely a clue to the fact that they are captured satellites, formed elsewhere before becoming moons of Uranus.

Only one moon in this irregular group, Margaret, has a **prograde orbit**, meaning it travels around Uranus in the same direction as the planet rotates. The remaining satellites have **retrograde orbits** instead, moving around the planet in the opposite direction to Uranus' spin. Both prograde and retrograde orbit directions are common for moons all over the Solar System and these directions can be a clue as to how a particular natural satellite formed. Prograde moons are more likely to be those that formed with their primary, as this would have established that same direction of movement for all the bodies. Retrograde orbits, however, often indicate moons that were formed from captured objects instead, as this process means that the object could have been gravitationally pulled in to orbit its primary from any angle and with any initial direction of travel. These factors mean that prograde moons are also more commonly closer to their primary than retrograde bodies, as it is rarer for captured objects to be able to recede as far inwards as pre-existing moons. The naming scheme for Jupiter's moons was set up to show the orbital direction: names ending in an 'a' indicate a prograde orbit for the satellite, whereas those with an 'e' at the end are in retrograde orbits. In the Himalia group of the irregular moons of Jupiter, for example, all the satellites have names ending in the letter 'a', while the names of those in the other three groups end with an 'e' instead.

Moons of Neptune

Neptune also has a mixture of prograde and retrograde moons. At the time of writing, Neptune is home to 14 moons which can be split, half and half, into two categories: the inner or regular group and the outer or irregular satellites. The regular moons orbit Neptune in regular circular, prograde orbit paths that are at the same angle as the planet's equator. Their tiny size, combined with Neptune's vast distance from Earth and the fact that *Voyager 2* is the only spacecraft to have visited the system so far (its closest approach to the planet occurred in August 1989) means that only the two largest regular moons, Proteus and Larissa, have been imaged in enough detail to determine their shapes and identify their surface features. At least eight times smaller than our Moon, these two satellites are still so small, however, that neither has enough mass to become spherical in shape; Larissa is an irregular rocky shape and Proteus is strangely box-like.

The regular moons orbit in and around Neptune's small, faint ring system, so one or two of them may be shepherd moons, clearing gaps in the planetary rings as we've observed around Saturn. The irregular moons travel around Neptune much further out, with the closest, Triton, orbiting at an average distance of around 354,759 km. The most distant two, Psamathe and Neso, have the largest orbits of any known natural satellites in the Solar System, orbiting their primary at an average distance of at least 125 times that between Earth and our Moon. It would take the equivalent of 25 Earth years to cover the distance of one complete orbit of Neptune!

The satellites in Neptune's irregular group have a mixture of prograde and retrograde orbits. They also travel around the planet on orbits with a range of different tilts and shapes. One of the most extreme orbits is that of one of the larger Neptunian moons, Nereid. The orbit path of this satellite is the least inclined and most eccentric of any known irregular satellite. **Orbital inclination** is how tilted an orbit path is compared to the equator of the object it is orbiting around. As Nereid has a very low orbital inclination, it orbits almost exactly above Neptune's equator. An orbit like this is sometimes called an **equatorial orbit**. **Eccentricity** is determined by an orbit's shape. The less eccentric the orbit, the more circular it is. This means that Nereid's incredibly eccentric orbit is shaped like an oval or ellipse, quite far from the perfect circle shape.

Nereid's unique orbit path is thought to be caused by the nearby irregular moon,

Triton. Triton was discovered just 17 days after Neptune itself by William Lassell, an English brewer by trade and keen amateur astronomer. He spotted Triton with his self-built, reflecting telescope, which he later donated to Royal Observatory Greenwich, in the 1880s. Lassell also discovered two of Uranus' moons, Ariel and Umbriel, and independently co-discovered the Saturnian moon, Hyperion.

Triton is Neptune's most massive moon, making up 99.5% of the mass orbiting the planet. It is also the biggest in terms of physical size with a diameter slightly larger than that of Pluto. The moon's composition is similar to Pluto too, with a rocky core making up about two thirds of the moon, surrounded by a subsurface ocean, then a water-ice crust layer, and topped with a nitrogen-rich, relatively young surface. Due to its size, composition and the fact that it is the only large moon in the Solar System to have a retrograde orbit, it is thought that Triton may have previously been a dwarf planet before being captured by Neptune's strong gravitational pull. This is because, even though it achieves hydrostatic equilibrium (it is spherical), there are too many other objects travelling around in this region that it would have been unlikely that a body of this mass could have used its gravity to clear its own clear and distinct path, a key aspect needed for something to be classed as a planet, rather than a dwarf planet.

This capture event is thought not only to be the cause of Nereid's strange orbital path, but also the reason for the formation of Neptune's inner moons. When Triton was first captured, its orbit would most likely have been far more eccentric than it is today, causing a lot of disruption to any other moons orbiting Neptune at the time and prompting them to collide with each other and break apart. Once the system had stabilised a little, the leftover debris and rubble would have clumped together to form some brand new inner moons – the ones we see today.

Along with Io and Enceladus, Triton is the third and final moon known to be volcanically active. Being one of the coldest places in our Solar System, with an average temperature of around -35 °C, it has cryovolcanoes (just like those on Enceladus), so it spews water and ammonia rather than liquid rock. One of the largest cryovolcanic features on Triton is Leviathan Patera, which seems to consist of a **caldera**, a region that was previously above a volcano's magma or cryomagma chamber but has collapsed since the eruption released the lava or cryolava, surrounded by a volcanic dome. With a diameter of around 2,000 km, if this is confirmed in the future to be a volcano, it will be one of the largest volcanoes in the whole Solar System. Triton is also the only place in the Solar System known to have cryolava lakes. These aren't permanent bodies of liquid, like those found on Earth or Titan, however, as the water and ammonia freeze after an eruption due to the cold temperatures.

Not all of the surface of this moon is young and refreshed though. Covering most of Triton's western half, there are rugged regions known as cantaloupe terrain because they resemble the skin of a cantaloupe melon. This feature is made of mostly dirty water ice and is unique to this moon.

Triton won't remain orbiting Neptune forever. Like Mars' moon Phobos, Triton's orbit is taking it on a slow spiral inward. In around 3.6 billion years, it will either collide with the planet or get torn apart by Neptune's gravity, perhaps forming a new planetary ring around it.

Miscellaneous Moons

We may have run out of planets in our Solar System, but there are still more moons to go!

For one thing, if you think back to our earlier definition, moons are objects travelling around primaries that are star-orbiting bodies. Other objects orbit stars, not just planets, so other types of bodies can have moons too. Let's start with dwarf planets, of which there are five official ones in our Solar System at the moment.

Moons of Pluto

The most famous dwarf planet is, of course, Pluto. A collision between this dwarf planet and another piece of space rock a few billion years ago is thought to have formed the Plutonian system of five moons (see Figure 14). The largest of these is called Charon. Some people say Charon with a 'sh' sound at the beginning, which links back to James Christy, the American astronomer who discovered the moon in 1978. He chose the name partly because it

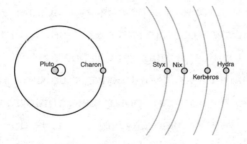

Figure 14: The dwarf planet Pluto has its own system of five moons. Charon is the largest.

sounded like his wife Charlene's nickname, 'Char'. Charon also echoes the name of the ferryman of the dead in Greek mythology, which is apt as Pluto was the ruler of the underworld. That, however, would be said in the original Greek with a 'k' sound at the beginning. As such, there is no official consensus for which pronunciation should be used.

Charon is the largest known moon in comparison to its parent body with a diameter just over half the size of Pluto's. Because of this, the two are close to being a **binary system**, a type of gravitational setup where two objects orbit each other, rather than being a primary and moon pair.

Charon has a distinctive reddish-brown cap in its northern polar region, informally known as Mordor Macula. This is due to the presence of tholins, organic compounds containing carbon that have formed as a result of exposure to radiation, either from the Sun or just background space.

Some scientists have speculated that these substances may be important for life: when combined with water, they could be the basis for simple, prebiotic chemistry – even life on Earth may have started with tholins. They can also be found on the moons Europa, Titan, Triton and possibly Rhea as well.

Another, more mysterious feature on Charon is Kubrick Mons, the largest in a particular range of mountains on the moon's surface. It was described as a 'mountain in a moat' when it was first imaged by NASA's *New Horizons* spacecraft, because, well, that's exactly what it looks like! It is a mountain peak surrounded by a sort of trench. It is not known how this unusual geological feature formed but one theory is that it could be a cryovolcano, like those found on Enceladus and Triton, and the surrounding 'moat' could be a strange kind of caldera, a region that has collapsed due to the ejection of the cryolava that was previously housed beneath it.

The names of the other four moons of Pluto (the Roman God of the underworld) are all linked to the Greek mythological underworld: Styx, named after the river that marks the boundary between Earth and the underworld; Nix, named after Nyx, the goddess of the night (the spelling was changed to the Egyptian version to avoid mixing it up with two asteroids of the same name); Kerberos (Cerberus to the Romans), the dog that guarded the underworld; and Hydra, named after the nine-headed serpent-like monster found there. These four moons are very reflective, with albedo values ranging between 0.5 and 0.8, due to their water icy surfaces, irregularly shaped, and tiny, with the largest, Nix, having a diameter of only around 50 km. They travel around Pluto at two to four times the distance that Charon does with very neat, circular, prograde orbit paths.

Other moons

The only other officially recognised dwarf planets in the Solar System – Ceres, Eris, Haumea and Makemake – have a handful of moons between them, though none of them have been observed in enough detail to be able to tell too much about them. Eris, the most massive of the dwarf planets has a single known satellite called Dysnomia, both of which were discovered in 2005 by astronomer Mike Brown and his team. Haumea has two moons that are named after Hawaiian goddesses, Hi'iaka and Namaka, as they were discovered from images taken by the W.M. Keck Observatory which is based in Hawaii. Makemake is home to one as yet unnamed moon, nicknamed MK2. Despite Makemake being the second brightest object in the Kuiper belt, a region of space similar to the asteroid belt but which lies beyond Neptune's orbit and contains objects made mostly of ices instead of asteroids, MK2 is extremely dark with an estimated albedo of only 0.04. One theory for this difference is that MK2 may not have enough gravity to hold on to icy, more reflective material enough to stop heat from the Sun allowing it to be lost to space.

There are a few other distant objects in the Solar System that some astronomers consider to be dwarf or minor planets, with moons of their own. Gonggong, for example, in the Kuiper belt, rotates particularly slowly on its axis compared to other trans-Neptunian objects, possibly due to tidal forces from its moon, Xiangliu. Varda, another Kuiper-belt object and dwarf-planet candidate, and its moon, Ilmarë, are named after deities from the works of Tolkien. Ilmarë was discovered in 2011 using images that were taken by the Hubble Space Telescope in 2009.

We have also discovered more than 150 moons orbiting around asteroids, objects made of rock and metal found mostly in the main asteroid belt which lies between the orbits of Mars and Jupiter. For instance, Elektra is currently the asteroid with the most moons – it's home to three. As all four bodies appear to be made of similar materials, it is thought that the satellites are pieces of Elektra that were broken off due to some kind of impact.

Moon-moons

But what about natural satellites themselves? Do any moons have moons? Well, theoretically, it could be possible. Objects that would orbit moons are known as subsatellites, submoons, or, my personal favourite, **moon-moons**.[3]

There are several artificial satellites currently orbiting our Moon, including

---

[3] Other terms suggested include moonitos, moonettes, and moooons!

NASA's Lunar Reconnaissance Orbiter (LRO), which has been studying the lunar surface since 2009. They could technically be classed as artificial moon-moons, or, perhaps a bit more accurately, subsatellites. These spacecraft can only stay in orbit for a (relatively) short time, however, before the gravity of either Earth or the Moon affects them too much and tries to pull them in downwards, towards the terrestrial or lunar surface. To compensate for this, their boosters have to fire every so often to keep them in a stable orbit path. The LRO, for example, was designed to make orbit-maintaining manoeuvres every 28 days. This demonstrates how difficult it is to find places where the gravities of the bodies involved would be able to balance out in a way that could keep a moon-moon in a constant orbit, rather than drifting off to orbit the primary (or star!) instead. Ideally you would need a massive moon orbiting a long way from its parent. We haven't yet identified a setup that matches this well enough to have any potential moon-moon locations.

Exomoons

We've covered all the known moons in our Solar System, but there are, of course, many more stars out there other than our Sun and so there are many more systems in which moons could be found. Planets orbiting around other stars are known as **exoplanets**, and, in turn, moons in other star systems are called **exomoons**. It's incredibly difficult to detect moons in our Solar System, let alone a far more distant one, which is why there are currently no confirmed exomoons. Over 20 promising candidates have been discovered, however. In July 2021, astronomers using the Atacama Large Millimetre/submillimeter Array (ALMA), an array of radio telescopes situated in the north Chilean desert, made the first clear detection of a moon-forming disc around an exoplanet known as PDS 70c, one of two planets that have been discovered orbiting around a star located around 370 light years away in the constellation of Centaurus. So exomoons are definitely out there.

But just how many moons do we think there are in the Universe? We have no reason to assume that our Solar System is special or unique with its 200+ moons. So, let's make some educated guesses. If we take an average for our Solar System, there are about 25 moons per planet. Let's round this down a bit to 20 in case we do have a few more than the typical system. In terms of planets, it is thought that, on average, there is one for every star out there. That means we can also take there to be 20 moons per star. In our own galaxy, the Milky Way, there are up to 400 billion stars. Based on our assumptions there could be up to 8,000 billion moons! And, of course, the Milky

Way is not the only galaxy out there. There are so many galaxies in the Universe that astronomers are still counting them and we are a long way off calculating a final number; a reasonable estimate, however, is thought to be somewhere in the region of 100 to 200 billion galaxies. Each one will be home to its own billions or even trillions of stars ... so, if our Solar System is anything to go by, there are certainly plenty more moons for us to discover!

Moons by Numbers

Moon	Primary	Year of Discovery	Average Equatorial Diameter (km)	Mass (lunar masses)	Average Orbital Distance (km)
Our Moon	Earth	Unknown	3475	1	384,400
Phobos	Mars	1877	22	0.00000015	9,376
Deimos	Mars	1877	12	0.00000002	23,458
Io	Jupiter	1610	3643	1.22	421,800
Europa	Jupiter	1610	3122	0.65	671,100
Ganymede	Jupiter	1610	5262	2.02	1,070,400
Callisto	Jupiter	1610	4821	1.46	1,882,700
Pan	Saturn	1990	28	0.00000007	133,580
Hyperion	Saturn	1848	270	0.000076	1,500,934
Mimas	Saturn	1789	396	0.00051	185,539
Enceladus	Saturn	1789	504	0.0015	238,037
Tethys	Saturn	1684	1066	0.0084	294,672

Dione	Saturn	1684	1123	0.015	377,415
Rhea	Saturn	1672	1529	0.031	527,068
Iapetus	Saturn	1671	1471	0.025	3,560,851
Titan	Saturn	1655	5149	1.83	1,221,865
Miranda	Uranus	1948	472	0.00090	129,900
Ariel	Uranus	1851	1158	0.018	190,900
Umbriel	Uranus	1851	1169	0.017	266,000
Titania	Uranus	1787	1578	0.047	436,300
Oberon	Uranus	1787	1523	0.039	583,500
Proteus	Neptune	1989	420	0.00069	117,646
Larissa	Neptune	1982	194	0.000067	73,548
Psamanthe	Neptune	2003	40	0.0000002	48,096,000
Neso	Neptune	2002	60	0.0000022	49,285,000
Nereid	Neptune	1949	340	0.00042	5,513,818
Triton	Neptune	1846	2707	0.29	354,759
Charon	Pluto	1978	1207	0.021	17,536

Glossary

anorthositic rock – a kind of igneous rock, formed when magma or lava cools inside existing rock. The lunar highlands are made of this.

albedo – a measure of how reflective a material or object is.

basaltic rock – a kind of dark, igneous rock formed from cooled magma or lava. Lunar *maria* are made of this.

binary system – a system of two bodies that orbit each other.

caldera – a pit or hollow found at the top of a volcano where the magma chamber has collapsed since eruption.

catenae – chains or lines of craters.

chaos terrain – region where plains, ridges and cracks appear muddled together.

chasmata – canyon-like features that are very long and deep with steep sides.

co-orbital moons – two or more moons that orbit around their primary at roughly the same distance out.

cryovolcano – a volcano powered by ices rather than magma.

eccentricity – how circular or oval-shaped a satellite's orbit is. More eccentric orbits are less round.

equatorial orbit – an orbit path that lies approximately over the primary's equator.

exomoon – a moon orbiting around an object in a system that is not our own Solar System.

exoplanet – a planet orbiting around a star that is not our own Sun.

Galilean moons – the four largest moons of Jupiter: Io, Europa, Ganymede and Callisto.

geysers – icy jets spraying up from a body's surface, usually associated with cryovolcanoes.

graben – valley bounded by two faults.

hydrostatic equilibrium – enough mass to overcome its own structural rigidity and gravitationally round itself.

infrared – a type of light or electromagnetic radiation that is invisible to the human eye as it has wavelengths longer than those of visible light.

isotopes – versions of the same element that contain different numbers of neutrons, therefore giving them different relative atomic masses but the same chemical properties.

leading hemisphere – the half of the moon that is facing the direction it is travelling in.

lenticulae – round, reddish spots found on the surface of Europa, Latin for 'freckles'.

linae – long, bright or dark lines on as body's surface.

lunar – relating to Earth's Moon.

lunar highlands – the lighter regions on our Moon's surface.

magnetosphere – a region surrounding a celestial object that is dominated by its magnetic field meaning that charged particles from the Sun are trapped inside it.

maria – dark patches on our Moon's surface formed from ancient, cooled lava.

moon – an astronomical object that orbits a primary (a planet or another star-orbiting object). Can also be referred to as a **natural satellite**.

moon-moon – an object orbiting around a moon, also referred to as a subsatellite or submoon.

multi-ring impact basin – a series of concentric craters.

natural satellite – another term for a moon.

orbital inclination – a measurement of how tilted a satellite's orbit path is compared to the equator of its parent.

orbital period – the length of time it takes an object to orbit around its primary, the equivalent of one year on that object.

orbital resonance – when two or more objects have a regular ratio between the amount of time they take to orbit around the same primary.

paterae – volcanic depressions such as those found on the moon Io.

penitentes – theoretical icy spikes thought to be around the equator of Europa.

phases – in the context of our Moon, a word used to describe the sequence different shapes the Moon appears to form from our perspective on Earth.

primary – the object around which the moon is orbiting.

prograde orbit – orbit of an object around its parent in the same direction as its parent's rotation.

retrograde orbit – orbit of an object around its parent in the opposite direction to its parent's rotation.

ring moonlet – mini moons orbiting inside the rings of a planet.

scarps – steep slopes that form when between two areas of different elevations that have been fractured apart.

shepherd moon – moon that orbit in and around the rings of a planet, helping to shape them with their gravity.

temporary moon – a natural satellite that is in an unstable orbit, so only remains travelling around its parent for a relatively small amount of time.

tholins – organic molecules that usually appear reddish in colour.

tidal heating – increase in heat of a body due to the push and pull from the gravity of nearby objects

tidal locking – the phenomena by which an object orbits its primary at the same speed as it rotates, causing only one side of the object to be visible to its primary.

trailing hemisphere – the half of the moon that is facing in the direction opposite to the direction of travel.

waning moon – the phase in which a moon's visible side is gradually decreasing, from full to new moon.

water ice – ice specifically made water rather than other chemicals, such as methane or ammonia.

waxing moon – the phase in which a moon's visible side is gradually increasing, from New to Full Moon.